Bump, Bump, Bump

Leslie Wood

Oxford University Press

Bump.
Out fell a clock.

Bump.
Out fell a vacuum cleaner.

Bump.
Out fell a piano.

Bump.
Out fell two chairs,
a violin, and a
bucket.

Bump.
Out fell a wheelbarrow
and a rake.

Bump.
Out fell a table.

Bump.
Out fell a carpet.

The van is empty!
How many things fell out?

Oxford University Press – Education
198 Madison Avenue, New York, New York 10016

Oxford New York
Athens Auckland Bangkok Bogota
Bombay Buenos Aires Calcutta Cape Town
Dar es Salaam Delhi Florence Hong Kong
Istanbul Karachi Kuala Lumpur Madras
Madrid Melbourne Mexico City Nairobi
Paris Singapore Taipei Tokyo Toronto

and associated companies in
Berlin Ibadan

Oxford is a registered trademark of Oxford University Press

ISBN 0 19 849016-X

Printed in Hong Kong